© Houghton Mifflin Harcourt Publishing Company • Cover Image Credits: (Ring-Necked Pheasant) ©Stephen Muskie/E+/Getty Images; (Field, Wisconsin) ©Image Studios/UpperCut Images/Getty Images

Houghton
Mifflin
Harcourt

Printed in the U.S.A.

ISBN 978-0-544-34235-4

19   0877   19

4500788160   B C D E F G

Dear Students and Families,

Welcome to **Go Math!**, Grade 5! In this exciting mathematics program, there are hands-on activities to do and real-world problems to solve. Best of all, you will write your ideas and answers right in your book. In **Go Math!**, writing and drawing on the pages helps you think deeply about what you are learning, and you will really understand math!

By the way, all of the pages in your **Go Math!** book are made using recycled paper. We wanted you to know that you can Go Green with **Go Math!**

Sincerely,

The Authors

Made in the United States
Text printed on 100% recycled paper

GO MATH!

Authors

Juli K. Dixon, Ph.D.
Professor, Mathematics Education
University of Central Florida
Orlando, Florida

Edward B. Burger, Ph.D.
President, Southwestern University
Georgetown, Texas

Steven J. Leinwand
Principal Research Analyst
American Institutes for
  Research (AIR)
Washington, D.C.

Contributor

Rena Petrello
Professor, Mathematics
Moorpark College
Moorpark, California

Matthew R. Larson, Ph.D.
K-12 Curriculum Specialist for
  Mathematics
Lincoln Public Schools
Lincoln, Nebraska

Martha E. Sandoval-Martinez
Math Instructor
El Camino College
Torrance, California

English Language Learners Consultant

Elizabeth Jiménez
CEO, GEMAS Consulting
Professional Expert on English
  Learner Education
Bilingual Education and
  Dual Language
Pomona, California

Fluency with Whole Numbers and Decimals

Critical Area Extending division to 2-digit divisors, integrating decimal fractions into the place value system and developing understanding of operations with decimals to hundredths, and developing fluency with whole number and decimal operations

GO DIGITAL

Go online! Your math lessons are interactive. Use *i*Tools, Animated Math Models, the Multimedia *e*Glossary, and more.

Chapter 5 Overview

In this chapter, you will explore and discover answers to the following **Essential Questions**:

• How can you solve decimal division problems?

• How is dividing with decimals similar to dividing with whole numbers?

• How can patterns, models, and drawings help you solve decimal division problems?

• How do you know where to place a decimal point in a quotient?

• How do you know the correct number of decimal places in a quotient?

Personal Math Trainer
Online Assessment and Intervention

CRITICAL AREA REVIEW PROJECT THE FORESTER: *www.thinkcentral.com*

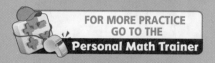

Practice and Homework

Lesson Check and Spiral Review in every lesson

FOR MORE PRACTICE GO TO THE **Personal Math Trainer**

Divide Decimals

✓ Show What You Know

Personal Math Trainer
Online Assessment and Intervention

Check your understanding of important skills.

Name _____

▶ **Division Facts** **Find the quotient.** (3.OA.C.7)

1. $6\overline{)24}$ = _____

2. $7\overline{)56}$ = _____

3. $18 \div 9$ = _____

4. $35 \div 5$ = _____

▶ **Estimate with 1-Digit Divisors** **Estimate the quotient.** (4.NBT.B.6)

5. $6\overline{)253}$

6. $4\overline{)1,165}$

7. $7\overline{)1,504}$

_____

_____

_____

▶ **Division** **Divide.** (5.NBT.B.6)

8. $34\overline{)785}$

9. $27\overline{)1,581}$

10. $41\overline{)4,592}$

Math in the Real World

Instead of telling Carmen her age, Sora gave her this clue. Find Sora's age.

Clue

My age is 10 more than one-tenth of one-tenth of one-tenth of 3,000.

Vocabulary Builder

▶ **Visualize It** •

Complete the bubble map using review words.

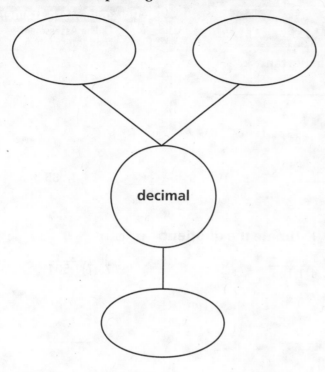

Review Words
compatible numbers
decimal
decimal point
dividend
divisor
equivalent fractions
estimate
exponent
hundredth
quotient
remainder
tenth

▶ **Understand Vocabulary** •

Complete the sentences using the review words.

1. A _____ is a symbol used to separate the ones place from the tenths place in decimal numbers.

2. Numbers that are easy to compute with mentally are called _____.

3. A _____ is one of ten equal parts.

4. A number with one or more digits to the right of the decimal point is called a _____.

5. The _____ is the number that is to be divided in a division problem.

6. A _____ is one of one hundred equal parts.

7. You can _____ to find a number that is close to the exact amount.

• **Interactive Student Edition**
• **Multimedia eGlossary**

decimal point (.)

punto decimal (.)

12

dividend

dividendo

18

divisor

divisor

19

equivalent fractions

fracciones equivalentes

22

estimate

estimación (s)
estimar (v)

23

exponent

exponente

26

quotient

cociente

57

remainder

residuo

59

The number that is to be divided in a division problem

Example: 36 ÷ 6 or 6)36

dividend

A symbol used to separate dollars from cents in money, and to separate the ones place and tenths place in a decimal

$1.65   4.324

decimal point

Fractions that name the same amount or part

Example: $\frac{1}{2}$ and $\frac{4}{8}$ are equivalent.

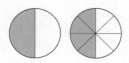

The number that divides the dividend

Example: 15 ÷ 3 or 3)15.

divisor

A number that shows how many times the base is used as a factor

exponent

Example: $10^3 = 10 \times 10 \times 10$

noun: A number close to an exact amount

verb: To find a number that is close to an exact amount

The amount left over when a number cannot be divided equally

Example:
```
      102 r2  ← remainder
   6)614
    −6
    ‾‾‾
     01
    −0
    ‾‾‾
     14
    −12
    ‾‾‾
      2  ← remainder
```

The number that results from dividing

Example: 8 ÷ 4 = 2

quotient

Picture It

For 3 to 4 players

Materials
- timer
- sketch pad

How to Play

1. Take turns to play.
2. To take a turn, choose a word from the Word Box. Do not say the word.
3. Set the timer for 1 minute.
4. Draw pictures and numbers to give clues about the word.
5. The first player to guess the word before time runs out gets 1 point. If he or she can use the word in a sentence, they get 1 more point. Then that player gets a turn choosing a word.
6. The first player to score 10 points wins.

Word Box

decimal point

dividend

divisor

equivalent fractions

estimate

exponent

quotient

remainder

The Write Way

Reflect

Choose one idea. Write about it.

- Write a story about a person who needs to estimate something.
- Tell what happens to the decimal point in this pattern.

  $763 \div 10^1$ **$763 \div 10^2$**   **$763 \div 10^3$**

- Explain equivalent fractions in your own words. Give an example.
- Tell how to solve this problem: $5\overline{)89.7}$ = _____.

Name _____

Division Patterns with Decimals

Essential Question How can patterns help you place the decimal point in a quotient?

Common Core · Number and Operations in Base Ten—5.NBT.A.2
MATHEMATICAL PRACTICES
MP5, MP6, MP7

Unlock the Problem

The Healthy Wheat Bakery uses 560 pounds of flour to make 1,000 loaves of bread. Each loaf contains the same amount of flour. How many pounds of flour does the bakery use in each loaf of bread?

You can use powers of ten to help you find quotients. Dividing by a power of 10 is the same as multiplying by 0.1, 0.01, or 0.001.

- Underline the sentence that tells you what you are trying to find.
- Circle the numbers you need to use.

One Way Use place-value patterns.

Divide. 560 ÷ 1,000

Look for a pattern in these products and quotients.

560 × 1 = 560	560 ÷ 1 = 560
560 × 0.1 = 56.0	560 ÷ 10 = 56.0
560 × 0.01 = 5.60	560 ÷ 100 = 5.60
560 × 0.001 = 0.560	560 ÷ 1,000 = 0.560

So, _____ pound of flour is used in each loaf of bread.

1. As you divide by increasing powers of 10, how does the position of the decimal point change in the quotients?

_____

Another Way Use exponents.

Divide. $560 \div 10^3$

Look for a pattern. $560 \div 10^0 = 560$

$560 \div 10^1 = 56.0$

$560 \div 10^2 = 5.60$

$560 \div 10^3 = $ _____

Remember
The zero power of 10 equals 1.
$10^0 = 1$
The first power of 10 equals 10.
$10^1 = 10$

2. Each divisor, or power of 10, is 10 times the divisor before it. How do the quotients compare?

_____

CONNECT Dividing by 10 is the same as multiplying by 0.1 or finding $\frac{1}{10}$ of a number.

🔢 Example

Liang used 25.5 pounds of tomatoes to make a large batch of salsa. He used one-tenth as many pounds of onions as pounds of tomatoes. He used one-hundredth as many pounds of green peppers as pounds of tomatoes. How many pounds of each ingredient did Liang use?

Tomatoes: 25.5 pounds

Onions: 25.5 pounds ÷ _____   **Green Peppers:** 25.5 pounds ÷ _____

Think: 25.5 ÷ 1 = _____   **Think:** _____ ÷ 1 = _____

25.5 ÷ 10 = _____   _____ ÷ 10 = _____

_____ ÷ 100 = _____

So, Liang used 25.5 pounds of tomatoes, _____ pounds of onions,

and _____ pound of green peppers.

Try This! **Complete the pattern.**

Ⓐ 32.6 ÷ 1 = _____

32.6 ÷ 10 = _____

32.6 ÷ 100 = _____

Ⓑ $50.2 ÷ 10^0$ = _____

$50.2 ÷ 10^1$ = _____

$50.2 ÷ 10^2$ = _____

 **Share and Show** MATH BOARD

Complete the pattern.

1. $456 ÷ 10^0 = 456$

$456 ÷ 10^1 = 45.6$

$456 ÷ 10^2 = 4.56$

$456 ÷ 10^3 =$ _____

Math Talk MATHEMATICAL PRACTICES ⑤

Use Patterns How can you determine where to place the decimal point in the quotient $47.3 ÷ 10^2$?

Think: The dividend is being divided by an increasing power of 10, so the decimal

point will move to the _____ one place for each increasing power of 10.

Name _____

Complete the pattern.

2. $225 \div 10^0 =$ _____

$225 \div 10^1 =$ _____

$225 \div 10^2 =$ _____

$225 \div 10^3 =$ _____

 3. $605 \div 10^0 =$ _____

$605 \div 10^1 =$ _____

$605 \div 10^2 =$ _____

$605 \div 10^3 =$ _____

 4. $74.3 \div 1 =$ _____

$74.3 \div 10 =$ _____

$74.3 \div 100 =$ _____

On Your Own

Math Talk MATHEMATICAL PRACTICES ⑦

Look for a Pattern What happens to the value of a number when you divide by 10, 100, or 1,000?

Complete the pattern.

5. $156 \div 1 =$ _____

$156 \div 10 =$ _____

$156 \div 100 =$ _____

$156 \div 1,000 =$ _____

6. $32 \div 1 =$ _____

$32 \div 10 =$ _____

$32 \div 100 =$ _____

$32 \div 1,000 =$ _____

7. $23 \div 10^0 =$ _____

$23 \div 10^1 =$ _____

$23 \div 10^2 =$ _____

$23 \div 10^3 =$ _____

8. $12.7 \div 1 =$ _____

$12.7 \div 10 =$ _____

$12.7 \div 100 =$ _____

9. $92.5 \div 10^0 =$ _____

$92.5 \div 10^1 =$ _____

$92.5 \div 10^2 =$ _____

10. $86.3 \div 10^0 =$ _____

$86.3 \div 10^1 =$ _____

$86.3 \div 10^2 =$ _____

MATHEMATICAL PRACTICE ⑦ **Look for a Pattern** **Algebra** Find the value of *n*.

11. $268 \div n = 0.268$

$n =$ _____

12. $n \div 10^2 = 0.123$

$n =$ _____

13. $n \div 10^1 = 4.6$

$n =$ _____

14. Go DEEPER Loretta is trying to build the largest taco in the world. She uses 2,000 pounds of ground beef, one-tenth as many pounds of cheese as beef, and one-hundredth as many pounds of lettuce as beef. How many pounds of lettuce and cheese combined did she use?

_____

Problem Solving • Applications

Use the table to solve 15–17.

15. **GO DEEPER** How much more cornmeal than flour does each muffin contain?

_____

16. **THINK SMARTER** If each muffin contains the same amount of sugar, how many kilograms of sugar, to the nearest thousandth, are in each corn muffin?

_____

Dry Ingredients for 1,000 Corn Muffins

Ingredient	Number of Kilograms
Cornmeal	150
Flour	110
Sugar	66.7
Baking powder	10
Salt	4.17

17. **MATHEMATICAL PRACTICE 5** **Use Patterns** The bakery decides to make only 100 corn muffins on Tuesday. How many kilograms of sugar will be needed?

_____

18. **WRITE** ⟩Math Explain how you know that the quotient $47.3 \div 10^1$ is equal to the product 47.3×0.1.

_____

_____

_____

_____

19. **THINK SMARTER** Use the numbers on the tiles to complete each number sentence.

$62.4 \div 10^0 =$ _____

$62.4 \div 10^1 =$ _____

$62.4 \div 10^2 =$ _____

| . | 0 | 2 |

| 4 | 6 |

Division Patterns with Decimals

Common Core
COMMON CORE STANDARD—5.NBT.A.2
Understand the place value system.

Complete the pattern.

1. $78.3 \div 1 =$ __78.3__

   $78.3 \div 10 =$ __7.83__

   $78.3 \div 100 =$ __0.783__

2. $179 \div 10^0 =$ _____

   $179 \div 10^1 =$ _____

   $179 \div 10^2 =$ _____

   $179 \div 10^3 =$ _____

3. $87.5 \div 10^0 =$ _____

   $87.5 \div 10^1 =$ _____

   $87.5 \div 10^2 =$ _____

4. $124 \div 1 =$ _____

   $124 \div 10 =$ _____

   $124 \div 100 =$ _____

   $124 \div 1,000 =$ _____

5. $18 \div 1 =$ _____

   $18 \div 10 =$ _____

   $18 \div 100 =$ _____

   $18 \div 1,000 =$ _____

6. $16 \div 10^0 =$ _____

   $16 \div 10^1 =$ _____

   $16 \div 10^2 =$ _____

   $16 \div 10^3 =$ _____

7. $51.8 \div 1 =$ _____

   $51.8 \div 10 =$ _____

   $51.8 \div 100 =$ _____

8. $49.3 \div 10^0 =$ _____

   $49.3 \div 10^1 =$ _____

   $49.3 \div 10^2 =$ _____

9. $32.4 \div 10^0 =$ _____

   $32.4 \div 10^1 =$ _____

   $32.4 \div 10^2 =$ _____

Problem Solving

10. The local café uses 510 cups of mixed vegetables to make 1,000 quarts of beef barley soup. Each quart of soup contains the same amount of vegetables. How many cups of vegetables are in each quart of soup?

11. The same café uses 18.5 cups of flour to make 100 servings of pancakes. How many cups of flour are in one serving of pancakes?

_____

_____

12. **WRITE** ▸*Math* Explain how to use a pattern to find $35.6 \div 10^2$.

_____

_____

Lesson Check (5.NBT.A.2)

1. The Statue of Liberty is 305.5 feet tall from the foundation of its pedestal to the top of its torch. Isla is building a model of the statue. The model will be one-hundredth times as tall as the actual statue. How tall will the model be?

2. Sue's teacher asked her to find $42.6 \div 10^2$. How many places and in what direction should Sue move the decimal point to get the correct quotient?

Spiral Review (5.NBT.A.1, 5.NBT.B.6, 5.NBT.B.7)

3. In the number 956,783,529, how does the value of the digit 5 in the ten millions place compare to the digit 5 in the hundreds place?

4. Taylor has $97.23 in her checking account. She uses her debit card to spend $29.74 and then deposits $118.08 into her account. What is Taylor's new balance?

5. At the bank, Brent exchanges $50 in bills for 50 one-dollar coins. The total mass of the coins is 405 grams. Estimate the mass of 1 one-dollar coin.

6. A commercial jetliner has 245 passenger seats. The seats are arranged in 49 equal rows. How many seats are in each row?

FOR MORE PRACTICE
GO TO THE
Personal Math Trainer

Name _____

Divide Decimals by Whole Numbers

Essential Question How can you use a model to divide a decimal by a whole number?

Common Core Number and Operations in Base Ten—5.NBT.B.7

MATHEMATICAL PRACTICES
MP1, MP2, MP5, MP6

Investigate

Materials ■ decimal models ■ color pencils

Angela has enough wood to make a picture frame with a perimeter of 2.4 meters. She wants the frame to be a square. What will be the length of each side of the frame?

A. Shade decimal models to show 2.4.

B. You need to share your model among _____ equal groups.

C. Since 2 wholes cannot be shared among 4 groups without regrouping, cut your model apart to show the tenths.

There are _____ tenths in 2.4.

Share the tenths equally among the 4 groups.

There are _____ ones and _____ tenths in each group.

Write a decimal for the amount in each group. _____

D. Use your model to complete the number sentence.

2.4 ÷ 4 = _____

So, the length of each side of the frame will be _____ meter.

Draw Conclusions

1. **MATHEMATICAL PRACTICE ⑤ Use a Concrete Model** Explain why you needed to cut apart the model in Step C.

_____

_____

2. Explain how your model would be different if the perimeter were 4.8 meters.

_____

_____

Make Connections

You can also use base-ten blocks to model division of a decimal by a whole number.

Materials ■ base-ten blocks

Kyle has a roll of ribbon 3.21 yards long. He cuts the ribbon into 3 equal lengths. How long is each piece of ribbon?

Divide. 3.21 ÷ 3

STEP 1

Use base-ten blocks to show 3.21.

Remember that a flat represents one, a long represents one tenth, and a small cube represents one hundredth.

There are _____ one(s), _____ tenth(s), and

_____ hundredth(s).

STEP 2 Share the ones.

Share the ones equally among 3 groups.

There is _____ one(s) shared in each group and _____ one(s) left over.

STEP 3 Share the tenths.

Two tenths cannot be shared among 3 groups without regrouping. Regroup the tenths by replacing them with hundredths.

There are _____ tenth(s) shared in each group and

_____ tenth(s) left over.

There are now _____ hundredth(s).

STEP 4 Share the hundredths.

Share the 21 hundredths equally among the 3 groups.

There are _____ hundredth(s) shared in each group

and _____ hundredth(s) left over.

So, each piece of ribbon is _____ yards long.

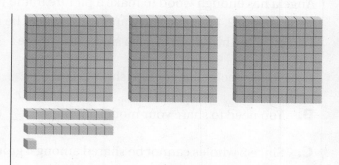

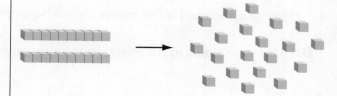

MATHEMATICAL PRACTICES 6

Explain why your answer makes sense.

Name _____

Use the model to complete the number sentence.

1. 1.6 ÷ 4 = _____

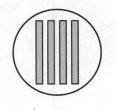 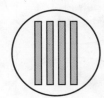

 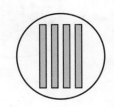

2. 3.42 ÷ 3 = _____

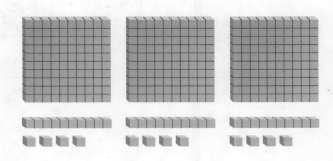

Divide. Use base-ten blocks.

3. 1.8 ÷ 3 = _____

4. 3.6 ÷ 4 = _____

5. 2.5 ÷ 5 = _____

6. 2.4 ÷ 8 = _____

7. 3.78 ÷ 3 = _____

8. 1.33 ÷ 7 = _____

9. 4.72 ÷ 4 = _____

10. 2.52 ÷ 9 = _____

11. 6.25 ÷ 5 = _____

Math Talk

MATHEMATICAL PRACTICES ①

Describe Relationships
Explain how you can use inverse operations to find 2.4 ÷ 4.

Problem Solving • Applications Real World

12. **THINK SMARTER** **What's the Error?**
Aida is making banners from a roll of paper that is 4.05 meters long. She will cut the paper into 3 equal lengths. She uses base-ten blocks to model how long each piece will be. Describe Aida's error.

 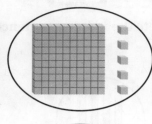 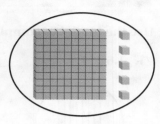

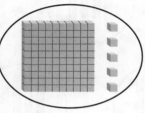

_____

_____

_____

_____

13. **GO DEEPER** Sam can ride his bike 4.5 kilometers in 9 minutes, and Amanda can ride her bike 3.6 kilometers in 6 minutes. Which rider might go farther in 1 minute?

_____

14. **MATHEMATICAL PRACTICE ②** **Use Reasoning** Explain how you can use inverse operations to find 1.8 ÷ 3.

_____

_____

15. **THINK SMARTER** Draw a model to show 4.8 ÷ 4 and solve.

4.8 ÷ 4 = _____

Divide Decimals by Whole Numbers

COMMON CORE STANDARD—5.NBT.B.7
Perform operations with multi-digit whole
numbers and with decimals to hundredths.

Use the model to complete the number sentence.

1. $1.2 \div 4 =$ ___0.3___

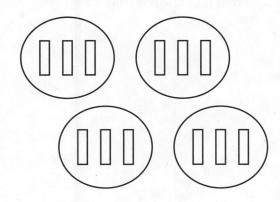

2. $3.69 \div 3 =$ _____

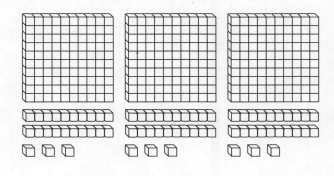

Divide. Use base-ten blocks.

3. $4.9 \div 7 =$ _____

4. $3.6 \div 9 =$ _____

5. $2.4 \div 8 =$ _____

6. $6.48 \div 4 =$ _____

7. $3.01 \div 7 =$ _____

8. $4.26 \div 3 =$ _____

Problem Solving Real World

9. In PE class, Carl runs a distance of 1.17 miles in 9 minutes. At that rate, how far does Carl run in one minute?

10. Marianne spends $9.45 on 5 greeting cards. Each card costs the same amount. What is the cost of one greeting card?

_____

_____

11. **WRITE** ▸ *Math* Explain how you can use base-ten blocks or other decimal models to find $3.15 \div 3$. Include pictures to support your explanation.

_____

_____

Lesson Check (5.NBT.B.7)

1. Write a division sentence that tells what the model represents.

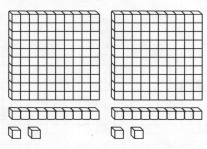

2. A bunch of 4 bananas contains a total of 5.92 grams of protein. Suppose each banana contains the same amount of protein. How much protein is in one banana?

_____

_____

Spiral Review (5.NBT.A.3b, 5.NBT.B.5, 5.NBT.B.6, 5.NBT.B.7)

3. At the deli, one pound of turkey costs $7.98. Mr. Epstein buys 3 pounds of turkey. How much will the turkey cost?

4. Mrs. Cho drives 45 miles in 1 hour. If her speed stays constant, how many hours will it take for her to drive 405 miles?

_____

_____

5. Write the following numbers in order from least to greatest.

1.23; 1.2; 2.31; 3.2

6. Over the weekend, Aiden spent 15 minutes on his math homework. He spent three times as much time on his science homework. How much time did Aiden spend on his science homework?

_____

_____

FOR MORE PRACTICE
GO TO THE
Personal Math Trainer

Estimate Quotients

Essential Question How can you estimate decimal quotients?

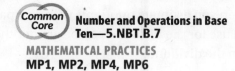

Common Core Number and Operations in Base Ten—5.NBT.B.7

MATHEMATICAL PRACTICES
MP1, MP2, MP4, MP6

 Unlock the Problem *Real World*

Carmen likes to ski. The ski resort where she goes to ski got 3.2 feet of snow during a 5-day period. The *average* daily snowfall for a given number of days is the quotient of the total amount of snow and the number of days. Estimate the average daily snowfall.

You can estimate decimal quotients by using compatible numbers. When choosing compatible numbers, you can look at the whole-number part of a decimal dividend or rename the decimal dividend as tenths or hundredths.

 Estimate. 3.2 ÷ 5

Carly and her friend Marco each find an estimate. Since the divisor is greater than the dividend, they both first rename 3.2 as tenths.

3.2 is _____ tenths.

CARLY'S ESTIMATE	MARCO'S ESTIMATE
30 tenths is close to 32 tenths and divides easily by 5. Use a basic fact to find 30 tenths ÷ 5.	35 tenths is close to 32 tenths and divides easily by 5. Use a basic fact to find 35 tenths ÷ 5.
30 tenths ÷ 5 is _____ tenths or _____.	35 tenths ÷ 5 is _____ tenths or _____.
So, the average daily snowfall is about _____ foot.	So, the average daily snowfall is about _____ foot.

1. **MATHEMATICAL PRACTICE ①** **Interpret a Result** Whose estimate do you think is closer to the exact quotient?

   Explain your reasoning. _____

   _____

2. Explain how you would rename the dividend in 29.7 ÷ 40 to choose compatible numbers and estimate the quotient.

   _____

   _____

Estimate with 2-Digit Divisors

When you estimate quotients with compatible numbers, the number you use for the dividend can be greater than the dividend or less than the dividend.

 **Example**

A group of 31 students is going to visit the museum. The total cost for the tickets is $144.15. About how much money will each student need to pay for a ticket?

Estimate. $144.15 ÷ 31

A **Use a whole number greater than the dividend.**

Use 30 for the divisor. Then find a number close to and greater than $144.15 that divides easily by 30.

$144.15 ÷ 31
  ↓   ↓
 $150 ÷ 30 = $ _____

So, each student will pay about $ _____ for a ticket.

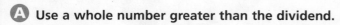

B **Use a whole number less than the dividend.**

Use 30 for the divisor. Then find a number close to and less than $144.15 that divides easily by 30.

$144.15 ÷ 31
  ↓   ↓
 $120 ÷ 30 = $ _____

So, each student will pay about $ _____ for a ticket.

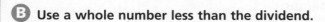

3. **MATHEMATICAL PRACTICE ②** **Use Reasoning** Which estimate do you think will be a better

estimate of the cost of a ticket? Explain your reasoning. _____

_____

_____

Share and Show

Use compatible numbers to estimate the quotient.

1. 28.8 ÷ 9

_____ ÷ _____ = _____

2. 393.5 ÷ 41

_____ ÷ _____ = _____

Name _____

Estimate the quotient.

3. $161.7 \div 7$

4. $17.9 \div 9$

5. $145.4 \div 21$

Math Talk

MATHEMATICAL PRACTICES ④

Interpret a Result Why might you want to find an estimate for a quotient?

On Your Own

Estimate the quotient.

6. $15.5 \div 4$

7. $394.8 \div 7$

8. $410.5 \div 18$

9. $72.1 \div 7$

10. $32.4 \div 52$

11. $\$134.42 \div 28$

12. **MATHEMATICAL PRACTICE ⑥** Shayne has a total of $135.22 to spend on souvenirs at the zoo. He wants to buy 9 of the same souvenir for his friends. Choose a method of estimation to find about how much Shayne can spend on each souvenir. **Explain** how you used the method to reach your estimation.

_____

_____

_____

13. **GO DEEPER** One week, Alaina ran 12 miles in 131.25 minutes. The next week, Alaina ran 12 miles in 119.5 minutes. If she ran a constant pace during each run, about how much faster did she run each mile in the second week than in the first week?

_____

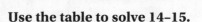

Problem Solving • Applications Real World

Use the table to solve 14–15.

14. **GO DEEPER** How does the estimate of the average daily snowfall for Wyoming's greatest 7-day snowfall compare to the estimate of the average daily snowfall for South Dakota's greatest 7-day snowfall?

_____

_____

15. **THINK SMARTER** The greatest monthly snowfall total in Alaska is 297.9 inches. This happened in February, 1953. Compare the daily average snowfall for February, 1953, with the average daily snowfall for Alaska's greatest 7-day snowfall. Use estimation.

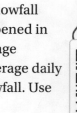

Greatest 7-Day Snowfall	
State	**Amount of Snow (in inches)**
Alaska	186.9
Wyoming	84.5
South Dakota	112.7

WRITE *Math* • **Show Your Work**

_____

_____

16. **WRITE** *Math* **What's the Error?** During a 3-hour storm, it snowed 2.5 inches. Jacob said that it snowed an average of about 8 inches per hour.

_____

_____

_____

_____

17. **THINK SMARTER** Juliette will cut a piece of string that is 45.1 feet long into 7 smaller pieces. Each of the 7 pieces will be the same length. Write a division sentence using compatible numbers to estimate the quotient.

_____

Estimate Quotients

COMMON CORE STANDARD—5.NBT.B.7
Perform operations with multi-digit whole numbers and with decimals to hundredths.

Use compatible numbers to estimate the quotient.

1. $19.7 \div 3$

$18 \div 3 = 6$

2. $394.6 \div 9$

3. $308.3 \div 15$

Estimate the quotient.

4. $63.5 \div 5$

5. $57.8 \div 81$

6. $172.6 \div 39$

7. $43.6 \div 8$

8. $2.8 \div 6$

9. $467.6 \div 8$

10. $209.3 \div 48$

11. $737.5 \div 9$

12. $256.1 \div 82$

Problem Solving · Real World

13. Taylor uses 645.6 gallons of water in 7 days. Suppose he uses the same amount of water each day. About how much water does Taylor use each day?

14. On a road trip, Sandy drives 368.7 miles. Her car uses a total of 18 gallons of gas. About how many miles per gallon does Sandy's car get?

15. **WRITE** ▸ *Math* Explain how to find an estimate for the quotient $3.4 \div 6$.

Lesson Check (5.NBT.B.7)

1. Terry bicycled 64.8 miles in 7 hours. What is the best estimate of the average number of miles she bicycled each hour?

2. What is the best estimate for the following quotient?

$$891.3 \div 28$$

_____

_____

Spiral Review (5.NBT.A.2, 5.NBT.A.3b, 5.NBT.B.7, 5.NF.B.3)

3. An object that weighs 1 pound on Earth weighs 1.19 pounds on Neptune. Suppose a dog weighs 9 pounds on Earth. How much would the same dog weigh on Neptune?

4. A bookstore orders 200 books. The books are packaged in boxes that hold 24 books each. All the boxes the bookstore receives are full, except one. How many boxes does the bookstore receive?

_____

_____

5. Tara has $2,000 in her savings account. David has one-tenth as much as Tara in his savings account. How much does David have in his savings account?

6. Which symbol makes the statement true? Write >, <, or =.

7.63 ◯ 7.629

_____

_____

FOR MORE PRACTICE
GO TO THE
Personal Math Trainer

Name _____

Division of Decimals by Whole Numbers

Essential Question How can you divide decimals by whole numbers?

Common Core Number and Operations in Base Ten—5.NBT.B.7
MATHEMATICAL PRACTICES
MP1, MP2, MP6

🔑 Unlock the Problem

In a swimming relay, each swimmer swims an equal part of the total distance. Brianna and 3 other swimmers won a relay in 5.68 minutes. What is the average time each relay team member swam?

> • How many swimmers are part of the relay team?
>
> _____

🔓 One Way Use place value.

MODEL

THINK AND RECORD

STEP 1 Divide the ones.

```
      1
  4)5.68
  -4
```

Divide. 5 ones ÷ 4

Multiply. 4 × 1 one

Subtract. 5 ones − 4 ones

Check. _____ one(s) cannot be shared among 4 groups without regrouping.

STEP 2 Divide the tenths.

```
      1
  4)5.68
  -4↓
  ___
  _
```

Divide. _____ tenths ÷ 4

Multiply. 4 × _____ tenths

Subtract. _____ tenths − _____ tenths

Check. _____ tenth(s) cannot be shared among 4 groups.

STEP 3 Divide the hundredths.

```
      1
  4)5.68
  -4
   16
  -16↓
  ___
  _
```

Divide. 8 hundredths ÷ 4

Multiply. 4 × _____ hundredths

Subtract. _____ hundredths − _____ hundredths

Check. _____ hundredth(s) cannot be shared among 4 groups.

Place the decimal point in the quotient to separate the ones and the tenths.

So, each girl swam an average of _____ minutes.

🔑 Another Way   Use an estimate.

Divide as you would with whole numbers.

Divide. $\$40.89 \div 47$

- Estimate the quotient. 4,000 hundredths ÷ 50 = 80 hundredths, or $0.80

- Divide the tenths.

- Divide the hundredths. When the remainder is zero and there are no more digits in the dividend, the division is complete.

- Use your estimate to place the decimal point. Place a zero to show there are no ones.

$$47\overline{)40.89}$$

So, $\$40.89 \div 47$ is _____ .

- **MATHEMATICAL PRACTICE ⑥** **Explain** how you used the estimate to place the decimal point in the quotient.

_____

_____

Try This! Divide. Use multiplication to check your work.

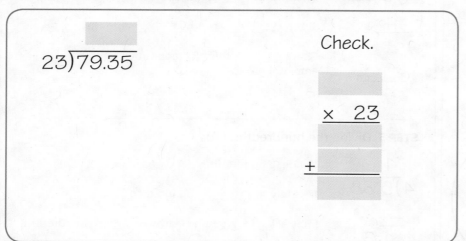

$$23\overline{)79.35}$$

Check.

$$\times \quad 23$$

$$+ \quad\underline{}$$

Share and Show   MATH BOARD ✏️

Write the quotient with the decimal point placed correctly.

1. $4.92 \div 2 = 246$ _____

2. $50.16 \div 38 = 132$ _____

Name _____

Divide.

3. $8)\overline{\$8.24}$

4. $3)\overline{2.52}$

5. $27)\overline{97.2}$

Math Talk

MATHEMATICAL PRACTICES ❶

Evaluate Reasonableness
How can you check that the decimal point is placed correctly in the quotient?

On Your Own

Practice: Copy and Solve Divide.

6. $3)\overline{\$7.71}$

7. $14)\overline{79.8}$

8. $33)\overline{25.41}$

9. $7)\overline{15.61}$

10. $14)\overline{137.2}$

11. $34)\overline{523.6}$

MATHEMATICAL PRACTICE ❷ Use Reasoning **Algebra** Write the unknown number for each ■.

12. $■ \div 5 = 1.21$

13. $46.8 \div 39 = ■$

14. $34.1 \div ■ = 22$

■ = _____

■ = _____

■ = _____

15. **THINK SMARTER** Mei runs 80.85 miles in 3 weeks. If she runs 5 days each week, what is the average distance she runs each day?

16. **GO DEEPER** Rob buys 6 tickets to the basketball game. He pays $8.50 for parking. His total cost is $40.54. What is the cost of each ticket?

Unlock the Problem Real World

17. **MATHEMATICAL PRACTICE ❶ Make Sense of Problems** The standard width of 8 lanes in swimming pools used for competitions is 21.92 meters. The standard width of 9 lanes is 21.96 meters. How much wider is each lane when there are 8 lanes than when there are 9 lanes?

a. What are you asked to find? _____

_____

b. What operations will you use to solve the problem? _____

_____

c. Show the steps you used to solve the problem.

d. Complete the sentences.

Each lane is _____ meters wide when there are 8 lanes.

Each lane is _____ meters wide when there are 9 lanes.

Since _____ − _____ = _____ , the

lanes are _____ meter(s) wider when there are 8 lanes than when there are 9 lanes.

18. **THINK SMARTER** Simon cut a pipe that was 5.75 feet long. Then he cut the pipe into 5 equal pieces. What is the length of each piece?

19. Jasmine uses 14.24 pounds of fruit for 16 servings of fruit salad. If each serving contains the same amount of fruit, how much fruit is in each serving?

Division of Decimals by Whole Numbers

Common
Core

COMMON CORE STANDARD—5.NBT.B.7
*Perform operations with multi-digit whole
numbers and with decimals to hundredths.*

Divide.

1.
```
      1.32
   7)9.24
    −7
    ──
     22
    −21
    ───
      14
     −14
     ───
       0
```

2. 6)5.04

3. 23)85.1

4. 36)86.4

5. 6)$6.48

6. 8)59.2

7. 5)2.35

8. 41)278.8

9. 19)$70.49

Problem Solving Real World

10. On Saturday, 12 friends go ice skating. Altogether, they pay $83.40 for admission. They share the cost equally. How much does each person pay?

_____

11. A team of 4 people participates in a 400-yard relay race. Each team member runs the same distance. The team completes the race in a total of 53.2 seconds. What is the average running time for each person?

_____

12. **WRITE** ▸*Math* Write a word problem involving money that requires dividing a decimal by a whole number. Include an estimate and a solution.

_____

_____

Lesson Check (5.NBT.A.2, 5.NBT.B.7)

1. Theresa pays $9.56 for 4 pounds of tomatoes. What is the cost of 1 pound of tomatoes?

2. Robert wrote the division problem below. What is the quotient?

$$13\overline{)83.2}$$

Spiral Review (5.OA.A.1, 5.NBT.A.2, 5.NBT.B.6, 5.NBT.B.7)

3. What is the value of the following expression?

$$2 \times \{6 + [12 \div (3 + 1)]\} - 1$$

4. Last month, Dory biked 11 times as many miles as Karly. Together they biked a total of 156 miles. How many miles did Dory bike last month?

5. Jin ran 15.2 miles over the weekend. He ran 6.75 miles on Saturday. How many miles did he run on Sunday?

6. A bakery used 475 pounds of apples to make 1,000 apple tarts. Each tart contains the same amount of apples. How many pounds of apples are used in each tart?

© Houghton Mifflin Harcourt Publishing Company

FOR MORE PRACTICE
GO TO THE
Personal Math Trainer

Name _____

Concepts and Skills

1. **Explain** how the position of the decimal point changes in a quotient as you divide by increasing powers of 10. (5.NBT.A.2)

   _____

2. **Explain** how you can use base-ten blocks to find $2.16 \div 3$. (5.NBT.B.7)

   _____

   _____

   _____

   _____

Complete the pattern. (5.NBT.A.2)

3. $223 \div 1 =$ _____

   $223 \div 10 =$ _____

   $223 \div 100 =$ _____

   $223 \div 1,000 =$ _____

4. $61 \div 1 =$ _____

   $61 \div 10 =$ _____

   $61 \div 100 =$ _____

   $61 \div 1,000 =$ _____

5. $57.4 \div 10^0 =$ _____

   $57.4 \div 10^1 =$ _____

   $57.4 \div 10^2 =$ _____

Estimate the quotient. (5.NBT.B.7)

6. $31.9 \div 4$

7. $6.1 \div 8$

8. $492.6 \div 48$

Divide. (5.NBT.B.7)

9. $5\overline{)4.35}$

10. $8\overline{)9.92}$

11. $61\overline{)207.4}$

© Houghton Mifflin Harcourt Publishing Company

12. The Westside Bakery uses 440 pounds of flour to make 1,000 loaves of bread. Each loaf contains the same amount of flour. How many pounds of flour are used in each loaf of bread? (5.NBT.A.2)

13. Elise pays $21.75 for 5 student tickets to the fair. What is the cost of each student ticket? (5.NBT.B.7)

14. Jason has a piece of wire that is 62.4 inches long. He cuts the wire into 3 equal pieces. Estimate the length of 1 piece of wire. (5.NBT.B.7)

15. GO DEEPER Elizabeth uses 23.25 ounces of granola and 10.5 ounces of raisins for 15 servings of trail mix. If each serving contains the same amount of trail mix, how much trail mix is in each serving? (5.NBT.B.7)

Name _____

Decimal Division

Essential Question How can you use a model to divide by a decimal?

Common Core — **Number and Operations in Base Ten—5.NBT.B.7**

MATHEMATICAL PRACTICES
MP2, MP4, MP5, MP6

Investigate

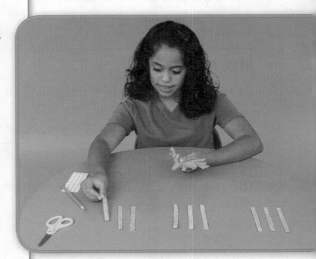

Materials ■ decimal models ■ color pencils

Lisa is making reusable shopping bags. She has 3.6 yards of fabric. She needs 0.3 yard of fabric for each bag. How many shopping bags can she make from the 3.6 yards of fabric?

A. Shade decimal models to show 3.6.

B. Cut apart your model to show the tenths. Separate the tenths into as many groups of 3 tenths as you can.

There are _____ groups of _____ tenths.

C. Use your model to complete the number sentence.

$3.6 \div 0.3 =$ _____

So, Lisa can make _____ shopping bags.

Draw Conclusions

1. Explain why you made each group equal to the divisor.

_____

_____

2.  **Represent a Problem** Identify the problem you would be modeling if each strip in the model represents 1.

_____

Remember
The divisor can tell the number of same-sized groups, or it can tell the number in each group.

3. MATHEMATICAL PRACTICE 5 **Communicate** Dennis has 2.7 yards of fabric to make bags that require 0.9 yard of fabric each. Describe a decimal model you can use to find how many bags he can make.

_____

_____

© Houghton Mifflin Harcourt Publishing Company

Make Connections

You can also use a model to divide by hundredths.

Materials ■ decimal models ■ color pencils

Julie has $1.75 in nickels. How many stacks of $0.25 can she make from $1.75?

STEP 1

Shade decimal models to show 1.75.

There are _____ one(s) and _____ hundredth(s).

STEP 2

Cut apart your model to show groups of 0.25.

There are _____ groups of _____ hundredths.

STEP 3

Use your model to complete the number sentence.

1.75 ÷ 0.25 = _____

So, Julie can make _____ stacks of $0.25 from $1.75.

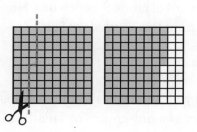

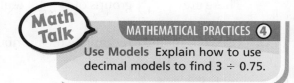

Math Talk

MATHEMATICAL PRACTICES ④

Use Models Explain how to use decimal models to find 3 ÷ 0.75.

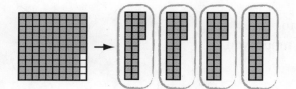

Use the model to complete the number sentence.

1. 1.2 ÷ 0.3 = _____

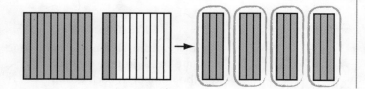

2. 0.45 ÷ 0.09 = _____

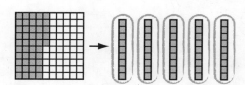

3. 0.96 ÷ 0.24 = _____

4. 1 ÷ 0.5 = _____

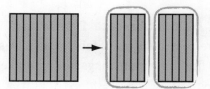

Name _____

Divide. Use decimal models.

5. $1.24 \div 0.62 = $ _____

6. $0.84 \div 0.14 = $ _____

7. $1.6 \div 0.4 = $ _____

Problem Solving • Applications

MATHEMATICAL PRACTICE (5) **Use Appropriate Tools** **Use the model to find the unknown value.**

8. $2.4 \div$ _____ $= 3$

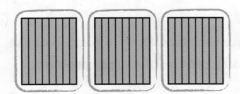

9. _____ $\div 0.32 = 4$

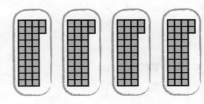

10. **THINK SMARTER** Make a model to find $0.6 \div 0.15$. Describe your model.

_____

_____

11. **MATHEMATICAL PRACTICE (6)** **Explain** using the model, what the equation represents in Exercise 9.

_____

_____

Personal Math Trainer

12. **THINK SMARTER +** Shade the model below and circle to show $1.8 \div 0.6$.

$$1.8 \div 0.6 = \boxed{}$$

THINK SMARTER **Pose a Problem**

13. Emilio buys 1.2 kilograms of grapes. He separates the grapes into packages that contain 0.3 kilogram of grapes each. How many packages of grapes does Emilio make?

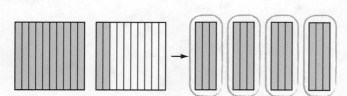

$1.2 \div 0.3 = 4$

Emilio made 4 packages of grapes.

Write a new problem using a different amount for the weight in each package. The amount should be a decimal with tenths. Use a total amount of 1.5 kilograms of grapes. Then use decimal models to solve your problem.

Pose a problem.

Solve your problem. Draw a picture of the model you used to solve your problem.

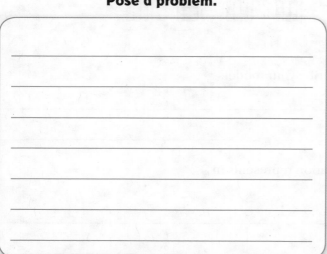

14. **GO DEEPER** Josie has 2.31 meters of blue ribbon that she wants to cut into 0.33-meter long pieces. She has 2.05 meters of red ribbon that she wants to cut into 0.41-meter long pieces. How many more pieces of blue ribbon than pieces of red ribbon will there be?

_____

Decimal Division

Common Core **COMMON CORE STANDARD—5.NBT.B.7**
Perform operations with multi-digit whole numbers and with decimals to hundredths.

Use the model to complete the number sentence.

1. 1.6 ÷ 0.4 = _____4_____

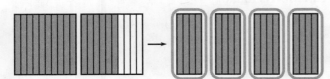

2. 0.36 ÷ 0.06 = _____

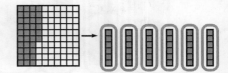

Divide. Use decimal models.

3. 2.8 ÷ 0.7 = _____

4. 0.40 ÷ 0.05 = _____

5. 0.45 ÷ 0.05 = _____

6. 1.62 ÷ 0.27 = _____

7. 0.56 ÷ 0.08 = _____

8. 1.8 ÷ 0.9 = _____

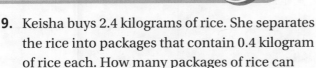

9. Keisha buys 2.4 kilograms of rice. She separates the rice into packages that contain 0.4 kilogram of rice each. How many packages of rice can Keisha make?

10. Leighton is making cloth headbands. She has 4.2 yards of cloth. She uses 0.2 yard of cloth for each headband. How many headbands can Leighton make from the length of cloth she has?

_____

_____

11. **WRITE** *Math* Write a word problem that involves dividing by a decimal. Include a picture of the solution using a model.

_____

_____

Lesson Check (5.NBT.B.7)

1. Write a number sentence that tells what the model represents.

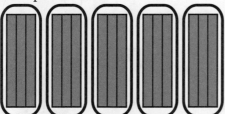

_____

2. Morris has 1.25 pounds of strawberries. He uses 0.25 pound of strawberries to make one serving. How many servings can Morris make?

_____

Spiral Review (5.NBT.B.5, 5.NBT.B.6, 5.NBT.B.7, 5.NF.B.3)

3. What property does the following equation show?

$$5 + 7 + 9 = 7 + 5 + 9$$

_____

4. An auditorium has 25 rows with 45 seats in each row. How many seats are there in all?

_____

5. Volunteers at an animal shelter divided 132 pounds of dry dog food equally into 16 bags. How many pounds of dog food did they put in each bag?

_____

6. At the movies, Aaron buys popcorn for $5.25 and a bottle of water for $2.50. He pays with a $10 bill. How much change should Aaron receive?

_____

FOR MORE PRACTICE
GO TO THE
Personal Math Trainer

Divide Decimals

Essential Question How can you place the decimal point in the quotient?

Common Core **Number and Operations in Base Ten—5.NBT.B.7**
Also 5.NBT.A.2
MATHEMATICAL PRACTICES
MP1, MP2, MP8

When you multiply both the divisor and the dividend by the same power of 10, the quotient stays the same.

dividend		divisor		dividend		divisor	
6	÷	3	= 2	120	÷	30	= 4
↓ × 10		↓ × 10		↓ × 0.1		↓ × 0.1	
60	÷	30	= 2	12	÷	3	= 4
↓ × 10		↓ × 10		↓ × 0.1		↓ × 0.1	
600	÷	300	= 2	1.2	÷	0.3	= 4

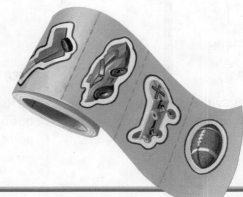

🔑 Unlock the Problem (Real World)

Matthew has $0.72. He wants to buy stickers that cost $0.08 each. How many stickers can he buy?

- Multiply both the dividend and the divisor by the power of 10 that makes the divisor a whole number. Then divide.

$$0.72 ÷ 0.08 = \boxed{}$$

↓ × 100 ↓ × 100

$$72 ÷ 8 = \boxed{}$$

So, Matthew can buy _____ stickers.

- What do you multiply hundredths by to get a whole number?

_____

1. **MATHEMATICAL PRACTICE ① Make Connections** Explain how you know that the quotient 0.72 ÷ 0.08 is equal to the quotient 72 ÷ 8.

_____

_____

Try This! Divide. 0.56 ÷ 0.7

- Multiply the divisor by a power of 10 to make it a whole number. Then multiply the dividend by the same power of 10.

  0.7 × _____ = _____

  0.56 × _____ = _____

- Divide.

$$0\,7.\overline{)5.6}$$

🔑 Example

Sherri hikes on the Pacific Coast trail. She plans to hike 3.72 miles. If she hikes at an average speed of 1.2 miles per hour, how long will she hike?

Divide. 3.72 ÷ 1.2

Estimate. _____

STEP 1	STEP 2	STEP 3
Multiply the divisor by a power of 10 to make it a whole number. Then, multiply the dividend by the same power of 10.	Write the decimal point in the quotient above the decimal point in the new dividend.	Divide.

STEP 1

1.2 × _____ = _____

3.72 × _____ = _____

STEP 2

$$12\overline{)37.2}$$

STEP 3

$$12\overline{)37.2}$$

So, Sherri will hike _____ hours.

2. **MATHEMATICAL PRACTICE 8** **Generalize** Describe what happens to the decimal point in the divisor and in the dividend when you multiply by 10.

_____

3. Explain how you could have used the estimate to place the decimal point.

_____

_____

Try This!

Divide. Check your answer.

$$0.14\overline{)1.96}$$

Multiply the divisor and the dividend by _____.

$$0.14$$
$$\times \underline{}$$

$$+ \underline{}$$

Name _____

Share and Show MATH BOARD

Copy and complete the pattern.

1. $45 \div 9 =$ _____

  $4.5 \div$ _____ $= 5$

  _____ $\div 0.09 = 5$

2. $175 \div 25 =$ _____

  $17.5 \div$ _____ $= 7$

  _____ $\div 0.25 = 7$

3. $164 \div 2 =$ _____

  $16.4 \div$ _____ $= 82$

  _____ $\div 0.02 = 82$

Divide.

✔ **4.** $1.6\overline{)9.6}$

5. $0.3\overline{)0.24}$

✔ **6.** $3.45 \div 1.5$

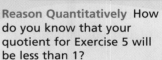

Reason Quantitatively How do you know that your quotient for Exercise 5 will be less than 1?

On Your Own

Divide.

7. $0.6\overline{)13.2}$

8. $0.3\overline{)0.9}$

9. $0.26\overline{)1.56}$

10. MATHEMATICAL PRACTICE ① Samuel has $0.96. He wants to buy erasers that cost $0.06 each. Describe how Samuel can find the number of erasers he can buy.

_____

_____

_____

11. GO DEEPER Penny makes 6 liters of applesauce. She saves 0.56 liter for dinner and puts the rest in jars. If each jar holds 0.68 liter, how many jars can she fill?

_____

_____

Problem Solving • Applications

Use the table to solve 12–16.

12. Connie paid $1.08 for pencils. How many pencils did she buy?

_____

13. Albert has $2.16. How many more pencils can he buy than markers?

_____

14. **GO DEEPER** How many erasers can Ayita buy for the same amount that she would pay for two notepads?

_____

15. **THINK SMARTER** Ramon paid $3.25 for notepads and $1.44 for markers. What is the total number of items he bought?

_____

16. Keisha has $2.00. She wants to buy 4 notepads. Does she have enough money? Explain your reasoning.

_____

_____

_____

17. **WRITE** ✏ *Math* **What's the Error?** Katie divided 4.25 by 0.25 and got a quotient of 0.17.

_____

_____

_____

18. **THINK SMARTER** Tara has a large box of dog treats that weighs 8.4 pounds. She uses the large box of dog treats to make smaller bags, each containing 0.6 pound of treats. How many smaller bags of dog treats can Tara make?

_____

Prices at School Store	
Item	**Price**
Eraser	$0.05
Marker	$0.36
Notepad	$0.65
Pencil	$0.12

WRITE ✏ *Math* • **Show Your Work**

Divide Decimals

Common Core **COMMON CORE STANDARD—5.NBT.B.7**
Perform operations with multi-digit whole numbers and with decimals to hundredths.

Divide.

1. $0.4\overline{)8.4}$

Multiply both 0.4 and 8.4 by 10 to make the divisor a whole number. Then divide.

$$\begin{array}{r} 21 \\ 4\overline{)84} \\ -8 \\ \hline 04 \\ -4 \\ \hline 0 \end{array}$$

2. $0.2\overline{)0.4}$

3. $0.07\overline{)1.68}$

4. $0.37\overline{)5.18}$

5. $0.4\overline{)10.4}$

6. $6.3 \div 0.7$

7. $1.52 \div 1.9$

8. $12.24 \div 0.34$

9. $10.81 \div 2.3$

Problem Solving Real World

10. At the market, grapes cost $0.85 per pound. Clarissa buys grapes and pays a total of $2.55. How many pounds of grapes does she buy?

11. Damon kayaks on a river near his home. He plans to kayak a total of 6.4 miles. Damon kayaks at an average speed of 1.6 miles per hour. How many hours will it take Damon to kayak the 6.4 miles?

_____

_____

12. **WRITE** ▸*Math* Write and solve a division problem involving decimals. Explain how you know where to place the decimal point in the quotient.

_____

_____

Lesson Check (5.NBT.A.2, 5.NBT.B.7)

1. Lee walked a total of 4.48 miles. He walks 1.4 miles each hour. How long did Lee walk?

2. Janelle has 3.6 yards of wire, which she wants to use to make bracelets. She needs 0.3 yard for each bracelet. Altogether, how many bracelets can Janelle make?

Spiral Review (5.NBT.A.2, 5.NBT.A.3b, 5.NBT.B.7)

3. Susie's teacher asks her to complete the multiplication problem below. What is the product?

$$\begin{array}{r} 0.3 \\ \times\ \ 3.7 \\ \hline \end{array}$$

4. At an Internet store, a laptop computer costs $724.99. At a local store, the same computer costs $879.95. What is the difference in prices?

5. Continue the pattern below. What is the quotient $75.8 \div 10^2$?

$$75.8 \div 10^0 = 75.8$$

$$75.8 \div 10^1 = 7.58$$

$$75.8 \div 10^2 = \underline{\hspace{2cm}}$$

6. Which symbol will make the following statement true? Write $>$, $<$, or $=$.

$$58.827 \bigcirc 58.91$$

FOR MORE PRACTICE GO TO THE Personal Math Trainer

Name _____

Write Zeros in the Dividend

Essential Question When do you write a zero in the dividend to find a quotient?

Common Core **Number and Operations in Base Ten—5.NBT.B.7**
Also 5.NF.B.3
MATHEMATICAL PRACTICES
MP2, MP3, MP5, MP6, MP8

CONNECT When decimals are divided, the dividend may not have enough digits for you to complete the division. In these cases, you can write zeros to the right of the last digit.

Unlock the Problem

The equivalent fractions show that writing zeros to the right of a decimal does not change the value.

$$90.8 = 90\frac{8 \times 10}{10 \times 10} = 90\frac{80}{100} = 90.80$$

During a fund-raising event, Adrian rode his bicycle 45.8 miles in 4 hours. Find his speed in miles per hour by dividing the distance by the time.

Divide. 45.8 ÷ 4   **Estimate.** 44 ÷ 4 = _____

STEP 1	STEP 2	STEP 3
Write the decimal point in the quotient above the decimal point in the dividend.	Divide the tens, ones, and tenths.	Write a zero in the dividend and continue dividing.

STEP 1:

$$4\overline{)45.8}$$

STEP 2:

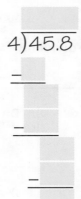

$$4\overline{)45.8}$$

STEP 3:

$$4\overline{)45.80}$$
$$-4$$
$$05$$
$$-4$$
$$18$$
$$-16 \downarrow$$

So, Adrian's speed was _____ miles per hour.

Math Talk

MATHEMATICAL PRACTICES ⑤
Use a Concrete Model How would you model this problem using base-ten blocks?

CONNECT When you divide whole numbers, you can show the amount that is left over by writing a remainder or a fraction. By writing zeros in the dividend, you can also show that amount as a decimal.

🔓 Example Write zeros in the dividend.

Divide. 372 ÷ 15

- Divide until you have an amount less than the divisor left over.

- Insert a decimal point and a zero at the end of the dividend.

- Place a decimal point in the quotient above the decimal point in the dividend.

- Continue dividing.

So, 372 ÷ 15 = _____ .

$$
\begin{array}{r}
24. \\
15\overline{)372.0} \\
-30 \\
\hline
72 \\
-60 \\
\hline
 \\
-
\end{array}
$$

- **MATHEMATICAL PRACTICE ⑥** Sarah has 78 ounces of rice. She puts an equal amount of rice in each of 12 bags. What amount of rice does she put in each bag? **Explain** how you would write the answer using a decimal.

_____

_____

Try This! Divide. Write a zero at the end of the dividend as needed.

Divide. 1.23 ÷ 0.06

$$
006.\overline{)123.} \qquad
\begin{array}{r}
20. \\
6\overline{)123.0} \\
-12 \\
\hline
03 \\
-0 \\
\hline
30 \\
-
\end{array}
$$

Divide. 10 ÷ 0.8

$$
08.\overline{)100.} \qquad
8.\overline{)100.}
$$

Name _____

Write the quotient with the decimal point placed correctly.

1. $5 \div 0.8 = 625$

2. $26.1 \div 6 = 435$

3. $0.42 \div 0.35 = 12$

4. $80 \div 50 = 16$

Divide.

5. $4\overline{)32.6}$

6. $1.2\overline{)9}$

✓ **7.** $15\overline{)42}$

✓ **8.** $0.14\overline{)0.91}$

On Your Own

Math Talk

MATHEMATICAL PRACTICES ⑧

Generalize Explain why you would write a zero in the dividend when dividing decimals.

Practice: Copy and Solve **Divide.**

9. $1.6\overline{)20}$

10. $15\overline{)4.8}$

11. $0.54\overline{)2.43}$

12. $28\overline{)98}$

13. $1.8 \div 12$

14. $3.5 \div 2.5$

15. $40 \div 16$

16. $2.24 \div 0.35$

17. **MATHEMATICAL PRACTICE ②** **Reason Quantitatively** Lana has a ribbon that is 2.2 meters long. She cuts the ribbon into 4 equal pieces to trim the edges of her bulletin board. What is the length of each piece of ribbon?

_____

18. **GO DEEPER** Hiro's family lives 448 kilometers from the beach. Each of the 5 adults drove the family van an equal distance to get to and from the beach. How far did each adult drive?

_____

Problem Solving • Applications

19. **GO DEEPER** Jerry takes trail mix on hikes. A package of dried apricots weighs 25.5 ounces. A package of sunflower seeds weighs 21 ounces. Jerry divides the apricots and seeds equally among 6 bags of trail mix. How many more ounces of apricots than seeds are in each bag?

_____

20. **THINK SMARTER** Amy has 3 pounds of raisins. She divides the raisins equally into 12 bags. How many pounds of raisins are in each bag? Tell how many zeros you had to write at the end of the dividend to solve.

_____

21. **MATHEMATICAL PRACTICE ③** **Compare Representations** Find $65 \div 4$. Write your answer using a remainder, a fraction, and a decimal. Then tell which form of the answer you prefer. Explain your choice.

_____

_____

_____

_____

22. **THINK SMARTER** For 22a–22d select Yes or No to indicate whether a zero must be written in the dividend to find the quotient.

22a.	$5.2 \div 8$	◯ Yes	◯ No
22b.	$3.63 \div 3$	◯ Yes	◯ No
22c.	$71.1 \div 0.9$	◯ Yes	◯ No
22d.	$2.25 \div 0.6$	◯ Yes	◯ No

Connect to Science

Rate of Speed Formula

The formula for velocity, or rate of speed, is $r = d \div t$, where r represents rate of speed, d represents distance, and t represents time. For example, if an object travels 12 feet in 10 seconds, you can find its rate of speed by using the formula.

$r = d \div t$

$r = 12 \div 10$

$r = 1.2$ feet per second

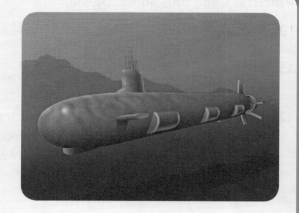

Use division and the formula for rate of speed to solve.

23. A car travels 168 miles in 3.2 hours. Find the car's rate of speed in miles per hour.

_____

24. A submarine travels 90 kilometers in 4 hours. Find the submarine's rate of speed in kilometers per hour.

Common Core **COMMON CORE STANDARD—5.NBT.B.7**
Perform operations with multi-digit whole numbers and with decimals to hundredths.

Divide.

1.
$$
\begin{array}{r}
3.95 \\
6\overline{)23.70} \\
-18 \\
\hline
57 \\
-54 \\
\hline
30 \\
-30 \\
\hline
0
\end{array}
$$

2. $25\overline{)405}$

3. $0.6\overline{)12.9}$

4. $0.8\overline{)30}$

5. $4\overline{)36.2}$

6. $35\overline{)97.3}$

7. $7.8 \div 15$

8. $49 \div 14$

9. $52.2 \div 12$

10. $5.16 \div 0.24$

11. $20.2 \div 4$

12. $138.4 \div 16$

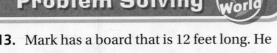

Problem Solving · Real World

13. Mark has a board that is 12 feet long. He cuts the board into 8 pieces that are the same length. How long is each piece?

14. Josh pays $7.59 for 2.2 pounds of ground turkey. What is the price per pound of the ground turkey?

_____   _____

15. **WRITE** ▸ *Math* Solve $14.2 \div 0.5$. Show your work and explain how you knew where to place the decimal point.

_____

_____

Lesson Check (5.NBT.B.7)

1. Tina divides 21.4 ounces of trail mix equally into 5 bags. How many ounces of trail mix are in each bag?

2. A slug crawls 5.62 meters in 0.4 hours. What is the slug's speed in meters per hour?

Spiral Review (5.NBT.A.2, 5.NBT.B.6, 5.NBT.B.7)

3. Suzy buys 35 pounds of rice. She divides it equally into 100 bags. How many pounds of rice does Suzy put in each bag?

4. Juliette spends $6.12 at the store. Morgan spends 3 times as much as Juliette. Jonah spends $4.29 more than Morgan. How much money does Jonah spend?

5. A concert sold out for 12 performances. Altogether, 8,208 tickets were sold. How many tickets were sold for each performance?

6. Jared has two dogs, Spot and Rover. Spot weighs 75.25 pounds. Rover weighs 48.8 pounds more than Spot. How much does Rover weigh?

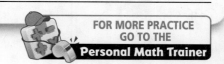

FOR MORE PRACTICE GO TO THE Personal Math Trainer

Name _____

Problem Solving • Decimal Operations

Essential Question How do you use the strategy *work backward* to solve multistep decimal problems?

 Common Core
Number and Operations in Base Ten—5.NBT.B.7
MATHEMATICAL PRACTICES
MP2, MP6, MP7

🔑 Unlock the Problem Real World

Carson spent $15.99 for 2 books and 3 pens. The books cost $4.95 each. The sales tax on the total purchase was $1.22. Carson also used a coupon for $0.50 off his purchase. If each pen had the same cost, how much did each pen cost?

Read the Problem

What do I need to find?	What information do I need to use?	How will I use the information?

Solve the Problem

- Make a flowchart to show the information. Then using inverse operations, work backward to solve.

Cost of 3 pens	plus →	Cost of 2 books	plus →	Amount of tax	minus →	Amount of Coupon	equals →	Total Spent
3 × cost of each pen	+	2 × ⬚	+	⬚	−	⬚	=	⬚

Total Spent	plus →	Amount of Coupon	minus →	Amount of tax	minus →	Cost of 2 books	equals →	Cost of 3 pens
⬚	+	⬚	−	⬚	−	⬚	=	⬚

- Divide the cost of 3 pens by 3 to find the cost of each pen.

_____ ÷ 3 = _____

Math Talk MATHEMATICAL PRACTICES ⑥

Explain why the amount of the coupon was added when you worked backward.

So, the cost of each pen was _____.

🔑 Try Another Problem

Last week, Vivian spent a total of $20.00. She spent $9.95 for tickets to the school fair, $5.95 for food, and the rest for 2 rings that were on sale at the school fair. If each ring had the same cost, how much did each ring cost?

Read the Problem

What do I need to find?	What information do I need to use?	How will I use the information?

Solve the Problem

So, the cost of each ring was _____.

MATHEMATICAL PRACTICES ❷

Use Reasoning How can you check your answer?

Name _____

1. Hector spent $36.75 for 2 DVDs that cost the same amount. The sales tax on his purchase was $2.15. Hector also used a coupon for $1.00 off his purchase. How much did each DVD cost?

   First, make a flowchart to show the information and show how you would work backward.

   Then, work backward to find the cost of 2 DVDs.

   _____

   Finally, find the cost of one DVD.

   _____

   So, each DVD costs _____.

2. **What if** Hector spent $40.15 for the DVDs, the sales tax was $2.55, and he didn't have a coupon? How much would each DVD cost?

   _____

3. Sophia spent $7.30 for school supplies. She spent $3.00 for a notebook and $1.75 for a pen. She also bought 3 large erasers. If each eraser had the same cost, how much did she spend for each eraser?

   _____

© Houghton Mifflin Harcourt Publishing Company

On Your Own

4. The change from a gift purchase was $3.90. Each of 6 students donated an equal amount for the gift. How much change should each student receive?

_____

5. **GO DEEPER** A mail truck picks up two boxes of mail from the post office. The total weight of the boxes is 32 pounds. One box is 8 pounds heavier than the other box. How much does each box weigh?

_____

6. **THINK SMARTER** Stacy buys 3 CDs in a set for $29.98. She saved $6.44 by buying the set instead of buying the individual CDs. If each CD costs the same amount, how much does each of the 3 CDs cost when purchased individually?

7. **MATHEMATICAL PRACTICE 7** **Look for a Pattern** A school cafeteria sold 1,280 slices of pizza the first week, 640 the second week, and 320 the third week. If this pattern continues, in what week will the cafeteria sell 40 slices? Explain how you got your answer.

_____

_____

_____

Personal Math Trainer

8. **THINK SMARTER +** Dawn spent $26.50, including sales tax on 4 books and 3 folders. The books cost $5.33 each and the total sales tax was $1.73. Fill in the table with the correct cost of each item.

Item	Cost
Cost of each book	
Cost of each folder	
Cost of sales tax	

Problem Solving • Decimal Operations

Common Core

COMMON CORE STANDARD—5.NBT.B.7
Perform operations with multi-digit whole numbers and with decimals to hundredths.

1. Lily spent $30.00 on a T-shirt, a sandwich, and 2 books. The T-shirt cost $8.95, and the sandwich cost $7.25. The books each cost the same amount. How much did each book cost?

(2 × cost of each book) + $8.95 + $7.25 = $30.00

$30.00 − $8.95 − $7.25 = (2 × cost of each book)

(2 × cost of each book) = $13.80
$13.80 ÷ 2 = $6.90

_____ $6.90

2. Meryl spends a total of $68.82 for 2 pairs of sneakers with the same cost. The sales tax is $5.32. Meryl also uses a coupon for $3.00 off her purchase. How much does each pair of sneakers cost?

_____

3. A 6-pack of undershirts costs $13.98. This is $3.96 less than the cost of buying 6 individual shirts. If each undershirt costs the same amount, how much does each undershirt cost when purchased individually?

_____

4. **WRITE** ▸*Math* Write a problem that can be solved using a flowchart and working backward. Then draw the flowchart and solve the problem.

_____

_____

_____

Lesson Check (5.NBT.B.7)

1. Joe spends $8 on lunch and $6.50 on dry cleaning. He also buys 2 shirts that each cost the same amount. Joe spends a total of $52. What is the cost of each shirt?

2. Tina uses a $50 gift certificate to buy a pair of pajamas for $17.97, a necklace for $25.49, and 3 pairs of socks that each cost the same amount. Tina has to pay $0.33 because the gift certificate does not cover the total cost of all the items. How much does each pair of socks cost?

Spiral Review (5.NBT.A.2, 5.NBT.A.3b, 5.NBT.B.7)

3. List the following numbers in order from least to greatest.

2.31, 2.13, 0.123, 3.12

4. Stephen wrote the problem 46.8 ÷ 0.5. What is the correct quotient?

5. Sarah, Juan, and Larry are on the track team. Last week, Sarah ran 8.25 miles, Juan ran 11.8 miles, and Larry ran 9.3 miles. How many miles did they run altogether?

6. On a fishing trip, Lucy and Ed caught one fish each. Ed's fish weighed 6.45 pounds. Lucy's fish weighed 1.6 times as much as Ed's fish. How much did Lucy's fish weigh?

FOR MORE PRACTICE GO TO THE
Personal Math Trainer

✓ Chapter 5 Review/Test

Personal Math Trainer
Online Assessment
and Intervention

1. Rita is hiking along a trail that is 13.7 miles long. So far she has hiked along one-tenth of the trail. How far has Rita hiked?

_____ miles

2. Use the numbers on the tiles to complete each number sentence. You can use a tile more than once or not at all.

$35.5 \div 10^0$ = [ ]

$35.5 \div 10$ = [ ]

$35.5 \div 10^2$ = [ ]

[.] [0] [3] [5]

3. **GO DEEPER** Tom and his brothers caught 100 fish on a weeklong fishing trip. The total weight of the fish was 235 pounds.

Part A

Write an expression that will find the weight of one fish. Assume that the weight of each fish is the same.

[ ]

Part B

What is the weight of one fish?

_____ pounds

Part C

Suppose the total weight of the fish caught stayed the same but instead of 100 fish caught during the weeklong fishing trip, only 10 fish were caught. How would the weight of each fish change? Explain.

[ ]

GO DIGITAL Assessment Options
Chapter Test

4. Draw a model to show $5.5 \div 5$.

$5.5 \div 5 = \boxed{}$

5. Emma, Brandy, and Damian will cut a rope that is 29.8 feet long into 3 jump ropes. Each of the 3 jump ropes will be the same length. Write a division sentence using compatible numbers to estimate the length of each rope.

6. Karl drove 617.3 miles. For each gallon of gas, the car can travel 41 miles. Select a reasonable estimate of the number of gallons of gas Karl used. Mark all that apply.

(A) 1.5 gallons

(B) 1.6 gallons

(C) 15 gallons

(D) 16 gallons

(E) 150 gallons

7. Donald bought a box of golf balls for $9.54. There were 18 golf balls in the box. About how much did each golf ball cost?

8. Luke cut down a tree that was 28.8 feet tall. Then he cut the tree into 6 equal pieces to take it away. What is the length of each piece?

_____ feet

9. Samantha is making some floral arrangements. The table shows the prices for one-half dozen of each type of flower.

Prices For $\frac{1}{2}$ Dozen Flowers	
Rose	$5.29
Carnation	$3.59
Tulip	$4.79

Part A

Samantha wants to buy 6 roses, 4 carnations, and 8 tulips. She estimates that she will spend about $14 on these flowers. Do you agree? Explain your answer.

Part B

Along with the flowers, Samantha bought 4 packages of glass beads and 2 vases. The vases cost $3.59 each and the total sales tax was $1.34. The total amount she paid was $28.50, including sales tax. Explain a strategy she could use to find the cost of 1 package of glass beads.

10. Les is sending 8 identical catalogs to one of his customers. If the package with the catalogs weighs 6.72 pounds, how much does each catalog weigh?

_____ pound(s)

11. Divide.

$$5\overline{)6.55}$$

12. Isabella is buying art supplies. The table shows the prices for the different items she buys.

Art Supplies	
Item	**Price**
Glass beads	$0.28 per ounce
Paint brush	$0.95
Poster board	$0.75
Jar of paint	$0.99

Part A

Isabella spends $2.25 on poster boards. How many poster boards does she buy?

_____ poster boards

Part B

Isabella spends $4.87 on paintbrushes and paint. How many of each item does she buy? Explain how you found your answer.

13. Shade the model and circle to show 1.4 ÷ 0.7.

1.4 ÷ 0.7 = ☐

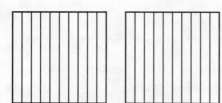

14. Tabitha bought peppers that cost $0.79 per pound. She paid $3.95 for the peppers. How many pounds of peppers did she buy? Show your work.

15. Hank has a large bag of trail mix that weighs 7.8 pounds. He uses the mix in the large bag to make bags each containing 0.6 pound of mix. How many bags containing 0.6 pound can be made?

_____ bags

16. Shareen walked a total of 9.52 miles in a walk-a-thon. If her average speed was 2.8 miles per hour, how long did it take Shareen to complete the walk?

_____ hours

17. For 17a–17c, choose Yes or No to indicate whether a zero must be written in the dividend to find the quotient.

17a. $1.4 \div 0.05$ ⚪ Yes ⚪ No

17b. $2.52 \div 0.6$ ⚪ Yes ⚪ No

17c. $2.61 \div 0.3$ ⚪ Yes ⚪ No

18. Lisandra made 22.8 quarts of split pea soup for her restaurant. She wants to put the same amount of soup into each of 15 containers. How much soup should Lisandra put into each container?

_____ quarts

19. Percy buys tomatoes that cost $0.58 per pound. He pays $2.03 for the tomatoes.

Part A

Percy estimates he bought 4 pounds of tomatoes. Is Percy's estimate reasonable? Explain.

[blank answer box]

Part B

How many pounds of tomatoes did Percy actually buy? Show your work.

[blank answer box]

20. Who drove the fastest? Select the correct answer.

(A) Harlin drove 363 miles in 6 hours.

(C) Shanna drove 500 miles in 8 hours.

(B) Kevin drove 435 miles in 7 hours.

(D) Hector drove 215 miles in 5 hours.

21. Maritza is buying a multipack of 3 pairs of socks for $25.98. She will save $6.39 by buying the multipack instead of buying 3 individual pairs of the same socks. If each pair of socks costs the same amount, how much does each pair of socks cost when bought individually? Show your work.

[blank answer box]

22. *THINK SMARTER +* Eric spent $22.00, including sales tax, on 2 jerseys and 3 pairs of socks. The jerseys cost $6.75 each and the total sales tax was $1.03. Fill in the table with the correct prices.

Personal Math Trainer

Item	Cost
Cost of each jersey	
Cost of each pair of socks	
Cost of sales tax	